BRITISH PIGS

Val Porter

SHIRE PUBLICATIONS

Published in Great Britain in 2009 by Shire Publications
Ltd, Midland House, West Way, Botley, Oxford OX2 0PH,
United Kingdom.
443 Park Avenue South, New York, NY 10016, USA.

E-mail: shire@shirebooks.co.uk   www.shirebooks.co.uk

A CIP catalogue record for this book is available from the
British Library.

Shire Library no. 340   •   ISBN-978 0 74780 763 6

Val Porter has asserted her right under the Copyright,
Designs and Patents Act, 1988, to be identified as the
author of this book.

Designed by Ken Vail Graphic Design, Cambridge, UK and
typeset in Perpetua and Gill Sans.
Printed in Malta by Gutenberg Press Ltd.

09 10 11 12 13   10 9 8 7 6 5 4 3 2 1

ACKNOWLEDGEMENTS
For their assistance in the preparation of this book, the
author is especially grateful to photographer Anna Oakford
and also to Caroline Benson (MERL), Richard Lutwyche,
Isabella Morton-Smith and Robin Quinnell.

Many of the illustrations are from the author's own
sources. Others are gratefully acknowledged as follows:

Abigail Chicken, page 40, Dani Clarke, page 47 (top and
middle); Josie Dew, page 51; Farmers Weekly Picture
Library, page 38 (bottom); Nick Hargrave, page 3; Zoe
Lindrop, page 48 (bottom); Richard Lutwyche, pages 8,
33, (middle and bottom); Museum of English Rural Life,
pages 4, 13, 17, 20 (both), 21, 24 (both), 25 (bottom), 28,
29 (top), 33 (top), 35 (top), and 50; Chris Murray, page
48 (top); National Pig Development Company, page 43
(bottom); Anna Oakford, front cover and pages 5
(bottom), 9 (both), 11, 16 (bottom), 26 (bottom), 32
(middle), 37 (bottom),
39 (top), 45 (bottom), and 62; Rare Breeds Survival Trust,
page 39 (bottom); Real Boar Company, page 15 (both),
and 55; Angus Stovold, page 22; Simon Tupper, page 42
(bottom); Sam Walton, page 46 (both); Philip Watts, page
26 (top), 27, 28 (bottom); Tony York, page 49 (both).

COVER IMAGE
Large Black farmyard sow and her litter (photograph by
Anna Oakford).

TITLE PAGE IMAGE
The adventurous golden red Tamworth is one of Britain's
most striking pig breeds.

CONTENTS PAGE IMAGE
British Saddlebacks.

Shire Publications is supporting the Woodland Trust, the UK's leading woodland conservation charity, by funding the dedication of trees.

CONTENTS

PIG BASICS

THE SUDDEN disappearance of pigs from the British landscape within living memory was remarkable. They have become no longer a familiar sight – to the extent that many people today are startled at the large size of a fully grown pig.

As recently as the 1950s, outdoor herds had been the norm on every pig farm. In addition, most smallholders reared a few outdoor pigs, and before the Second World War many a cottage dweller still kept the traditional family pig at the end of the garden, fattening it on household scraps, orchard windfalls and vegetable-patch waste, and regarding it almost as a member of the household – apart from its destiny as home-killed, home-cured meat for the table.

In the post-war decades pig farming changed radically and many pigs became hidden from public view, raised on a large scale within purpose-made buildings where highly bred and efficient commercial animals were intensively reared to supply consumers with affordable bacon, pork and sausages.

Wallowing is a pig's delight.

A portable farrowing hut for outdoor pigs in the late 1930s.

In 2008 about 3 per cent of British pigs were defined as 'free-range', living in fields with their mother until slaughter, and 18 per cent were described as 'outdoor-reared': the piglets were born outside and lived outside for 4–12 weeks before being moved into open barns for fattening. Fifty-nine per cent of UK pig production was from indoor-reared piglets, living in groups of 80–100 (often on slatted concrete) until slaughter.

5

Pig improvement companies evolved, creating new white hybrids and lines that were able to thrive in these intensive conditions. The companies became masters of the intricacies of pig genetics, each breeding many thousands of young pigs annually for sale to pig-rearing farmers with indoor enterprises of their own. Of the five million pigs in the United Kingdom in the opening years of the twenty-first century, about 10 per cent were breeding sows, most of them kept indoors, and about 70 per cent of their offspring were reared intensively indoors – most of them crossbreds based on Large White and Landrace. The young are slaughtered for meat at between four and seven months of age.

However, the situation is changing. New legislation and different attitudes to farm animal welfare have meant that increasing numbers of breeding sows are now being kept in more extensive systems out of doors. Breeding programmes reflect the change, and the search is on for genes that can make pigs hardier again – as they used to be. Some of the traditional minor British breeds might be just what the industry needs for the future.

Wiltshire bacon curing on the farm in the traditional way.

Writing in 1823, William Cobbett advised cottagers on the keeping of pigs, but, he said, 'these are animals not to be ventured on without due consideration

as to the means of *feeding* them; for, a starved pig is a great deal worse than none at all. You cannot make bacon, as you can milk, merely out of the garden. There must be *something more*. A couple of flitches of bacon are worth fifty thousand methodist sermons and religious tracts. The sight of them upon the rack tends more to keep a man from poaching and stealing than whole volumes of penal statutes, though assisted by the terrors of the hulks and the gibbet. They are great softeners of the temper and promoters of domestic harmony. They are a great blessing ...'

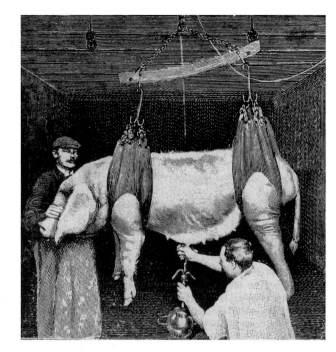

The lost art of milking a pig ...

Cobbett did not discuss breeds of pig: he simply advised the cottager to let the farmer do the breeding and then to buy from the farmer a ready-weaned pig, which, 'at *four months old*, ... if he be in good condition, he will eat any thing that an old hog will eat. He will graze, eat cabbage leaves and almost the stumps, Swedish turnip tops or roots; and such things, with a little wash, will keep him along in very good growing order.' Such a weanling would be a year old by its killing time, but 'If a hog be more than a year old, he is the better for it. The flesh is more solid and more nutritious than that of a young hog ... The pork or bacon of young hogs, even if fatted on corn, is very apt to *boil out*, as they call it; that is to say, come out of the pot smaller in bulk than it goes in.' Around Christmas was a good time to kill. 'To kill a hog nicely is so much of a profession, that it is better to pay a shilling for having it done, than to stab and hack and tear the carcass about.'

Cobbett went into considerable detail about how to deal with the fresh carcass, burning off the hair, scraping the skin, taking out the innards ('and if the wife be not a slattern, here, in the mere offal, in the mere garbage, there is food, and delicate food too, for a large family for a week; and hogs' puddings for the children, and some for neighbours' children who come to play with them'). The next day the butcher would cut up the hog: Cobbett gives a long list of the different cuts of meat that the family would achieve, and the supply of lard that would be available, even before the remaining sides, or flitches, were set aside to be cured as bacon in the home salting-trough – enough to last until the following Christmas. Ideally the flitches (or preferably just the hams) could be smoked in the cottage's inglenook bacon-loft. And in the meantime the cottage pig had been merrily supplying a considerable quantity of dung to enrich the vegetable patch, as well as turning over the soil in its rootling, making it easier for the cottager to dig.

It is easy to like pigs. They are friendly, sociable, talkative animals, highly intelligent and trainable, and easily pleased by human company. Their physiology is far closer to that of humans than that of other livestock. For example, pig's milk and the pig's digestive system are similar to those of humans.

Pigs are omnivores and will try almost anything edible. Though mainly vegetarian, eating plant roots, tree mast, berries, fruit, vegetables and grain, they will also rootle for worms, grubs and the occasional small vertebrates and carrion. They are great opportunists and well able to look after themselves.

Most domestic pigs have curly tails, though wild pigs and many Asian domestic pigs have straight tails. The most obvious characteristics that differentiate the various pig breeds are the carriage of the ears, the shape of the face, the colour of the skin and coat, and the general conformation – body shape and size. The table gives a broad guide to the more obvious differences between the breeds described later.

An Essex and a Large Black enjoying a good rootle.

BREED IDENTIFICATION CHECKLIST
Traditional British breeds at a glance

Breed	Colours	Features
Berkshire	black with white touches (feet, face, tassel)	prick ears; dished face
British Landrace	white	lop ears
British Lop	white	lop ears
British Saddleback	black with white belt and forelegs (also white hind legs for Essex)	lop ears
Gloucestershire Old Spots	white with defined black spot(s)	lop ears
Large Black	mealy black all over; black skin	lop ears
Large White	white	prick ears
Middle White	white	prick ears; dished face with snubbed snout
Oxford Sandy and Black	sandy (pale to rust) with random black blotches; white blaze, feet and tassel	lop or semi-lop ears
Tamworth	golden-red	prick ears; long snout (slightly dished)
Welsh	white	lop ears meeting at tips

Newcomers

Breed	Colours	Features
British Duroc (USA)	cherry red/auburn	ears pointing forward, not erect; dished face
Hampshire (USA)	black with white belt	prick ears or slightly forward
Iron Age (UK)	dark; piglets striped	prick ears; long snout
Kunekune (New Zealand)	ginger, brown, cream, black or combinations	prick or floppy ears; dished face
Mangalitza (Hungary)	blonde, red or 'swallow-bellied' (black with white underline); skin grey/black	dense curly coat; lop ears (upright in red)
Meishan (China)	black with white feet	heavy drooping ears; wrinkled skin and face
Vietnamese Pot-bellied	black	prick ears; wrinkled skin and snout; pot belly

The Rare Breeds Survival Trust in January 2009 classified pig breeds as follows:

Endangered: British Lop; Middle White.

Vulnerable: Berkshire; Large Black; Tamworth; Welsh.

At risk: British Saddleback.

Minority: Gloucestershire Old Spots.

Other native breeds: Large White.

Prick ears and a
good long snout
for rootling.

Prick ears and a
squashed face
typical of the
Middle White
breed.

Ears range from 'prick' (upright) to 'lop' (falling over the face), with a range of semi-prick and semi-lop in between. Snouts range from long and tapered, typically seen in the Wild Boar, to dished, snubbed and squashed in breeds influenced long ago by Asian imports.

Skin and coat colours include white, black and shades of red, brown or sandy, either all the same colour or with patches or spots of other colours. Coats vary in thickness and texture: some pigs are sparsely haired, some lushly; some have fine hair, others thick and wavy hair, or coarser bristles. Some even have fleece-like curly coats. Oriental pigs often have wrinkled skins, like a Shar Pei puppy, and their wrinkled snouts are apparent in Britain's Middle White breed.

These obvious visual differences are the badges that define each breed. Pigs can also differ in the roles for which they are bred: to produce rashers of lean bacon, for example, or to produce pork. Today commercial breeds are less likely to be specialists, and the British pig industry has depended very heavily on white breeds, above all the Large White, the

Landrace and the Welsh, though coloured sires are increasingly used to make more robust litters for outdoor rearing.

These three commercial whites apart, all of the other British breeds are now minor or very rare, despite often having been exported in considerable numbers to many parts of the world (there is British blood in many breeds in Europe, Australasia and North America in particular). Most of today's minor British breeds number only a few hundred each at best, nationwide, and come under the watchful eye of the Rare Breeds Survival Trust. The coloured breeds that were once ubiquitous have almost vanished. There were good commercial reasons for concentrating on the whites, and the pig industry in Britain has built up a remarkably successful record since the first of the big pig-breeding companies was formed in the early 1960s. But efficiency is not everything, and many people are saddened at the loss of

Lop ears falling over the face in the Large Black.

Right: Smallholders experiment with crossing breeds to produce meat for the home table and for farmers' markets, achieving a variety of coat patterns and types. This free-range herd is based on a Large Black boar and assorted sow breeds, including Gloucestershire Old Spots, Tamworth and Large White.

Below: English market scene in the 1940s, with a good assortment of breeds.

colour in the world of pigs. Some of the character seems to have been washed away with the colour.

Once upon a time every county boasted its own breed of pig, and even the occasional town, village, duke or prince did too. Today that close local identity has disappeared, and a host of once familiar breeds are extinct. Something of Britain's heritage has vanished with them.

A pandemonium of pigs at the Royal Show, 1997, including Tamworth, Gloucestershire Old Spots, Middle White and Large Black.

WILD BOAR AND DOMESTICATION

T HE ANCESTOR of the British domestic pig is the native Eurasian Wild Boar (*Sus scrofa*), which used to range all over Europe and Asia, in climates varying from typically British to Siberian and tropical. The domestic pig, under the skin, is not so different from the Wild Boar.

The Wild Boar's natural environment is woodland, scrubland or steppe, with the luxury of a muddy place to wallow on occasion. The animals live in a matriarchal society based on one or more females and their daughters; the males tend to live separately. Sows might have up to fifteen striped piglets in a litter, but they have only a dozen teats and in most cases each piglet demands its own teat. Wild Boar, like other wild pig species, are vocal and communicate constantly within the family group with grunts and squeaks. They feed mainly on plants (foliage, roots, fruit, bulbs) and fungi but will also eat earthworms. Their ability to rootle into plant debris and moist soil is legendary.

The grizzly coat of the Wild Boar comprises coarse dark brownish-grey bristles, paler on the face and undersides but darker on the ears, snout, lower legs and tail. Wild Boar are monomorphic in their coat colour: they show none of the patterns and colour variations seen in domesticated pigs, except that the piglets are born striped. The tail is bristly and straight, perhaps 12–16 inches (30–40 cm) in length. The snout is long, straight and tapering; the ears are erect and the eyes are small (the senses of smell and hearing are better than eyesight). They have cloven hoofs and tusks, which are actually protruding lower canine teeth, larger in the male and useful slashing weapons when fighting.

The Wild Boar varies in size over its enormous natural range. The largest are in the north and west, the smallest in the south and east, and so Britain's native Wild Boar was among the largest. European Wild Boar males can measure as much as 6 feet (182 cm) long, twice the length of some of those in south-east Asia. In the wild these ancestral swine can live for twenty years or so, if they can escape the hunter. There are still free-ranging Wild Boar in many European countries, though the species has been hunted (for meat and

Left and below: Free-range imported continental Wild Boar being reared for meat in the Cotswolds. Wild Boar piglets are born with stripes (excellent camouflage in their natural environment).

Wild Boar still roamed freely in central and northern Europe in the nineteenth century. The sows bred only once a year and seldom had more than half-a-dozen young.

for sport) for about forty thousand years. The last native Wild Boar in Britain was killed during the early seventeenth century.

Hunting, ironically, was a first step towards the domestication of the Wild Boar. The piglets of a slaughtered female could easily be reared in captivity. Being scavengers by nature, Wild Boar were often in the vicinity of humans anyway, looking for their leavings. The earliest domestication was probably in south-west Asia some nine or ten thousand years ago. Immigrant domesticated pigs reached Britain about five thousand years ago.

Domesticated pigs in medieval Britain contrasted with the Wild Boar in that they tended to have lop ears rather than pricked and their tails tended

Ginger-and-white offspring from a commercial white sow bred to a 'Wild Boar x Berkshire' boar. First crosses between British breeds and Wild Boar can have a variety of coat colours and patterns but the piglets are born with wild-type stripes.

The typical Old English pig, here a sow from the Midlands (reproduced by Professor David Low in 1842). In the 1930s the *Pig Breeders' Annual* rated the Old English type more highly than all the breeds subsequently influenced by imported Chinese pigs, and pointed out the similarities of the Old English with the ancient pigs depicted on a relief at the Forum in Rome.

to be curly rather than straight. For many centuries the typical **Old English** hog had big slouching ears hanging over its eyes (said to make the animal more docile as it could not see where it was going), a narrow razor back, low shoulders, flat slab sides, a long rootling snout, and long strong legs so that it could range widely in its foraging and walk long distances to market. It was a large-framed, hairy animal, slow to grow, and usually a dirty yellow-brown in colour, with or without spots or belts of another colour. This was the common 'Celtic' domestic pig of northern Europe.

In 1870 Joseph Harris, writing at Moreton Farm, Rochester, in the state of New York, reproduced an etching of the 'Original Old English Pig', differentiated from the Wild Boar in that its tail was curled and its ears hung down. Harris said that it showed 'a decided improvement in form over the Wild Boar. It has shorter legs, shorter head and snout, heavier cheeks, a straighter and broader back, and larger hams. It will weigh more, in proportion to size, and afford more meat and less offal than the wild hog.'

This Old English hog was a gleaner. Under the eye of the village swineherd, it found what food it could on common pasture, on stubble and in the woodlands (this was the ancient practice of pannage, or turning out the pigs to rootle for acorns and beech mast in autumn). It fed cheaply, but it took at least sixteen months to reach slaughter weight, and that became a drawback, especially when the Industrial Revolution caused people to crowd into towns and cities and so become reliant on other people to produce their food in large quantities.

Joseph Harris's illustration of the original 'Old English' pig, said to be an improvement on the Old Irish 'Greyhound' type achieved merely by 'regular feeding and judicious selection'.

Harris also illustrated and described the **Old Irish Pig**, which he said was an intermediate form between the wild and the domestic animal. He quoted from Richardson (from whose work the picture had been borrowed): 'These are tall, long-legged, bony, heavy-eared, coarse haired animals; their throats furnished with pendulous wattles, and by no means possessing half so much the appearance of domestic swine as they do of the wild boar, the great original of the race. In Ireland, the old gaunt race of hogs has, for many years past, been gradually wearing away, and is now, perhaps, wholly confined to the western parts of the country, especially Galway. These swine are remarkably active, and will clear a five-barred gate as well as any hunter; on this account they should, if it is desirable to keep them, be kept in well-fenced inclosures.'

Harris believed that the Old English pig he had described showed that 'great improvement can be made merely by regular feeding and judicious selection; but it must be remembered that probably it took hundreds of generations to effect the change indicated in the engravings. ... the fact remains that, centuries after the wild pigs had generally disappeared from the Island, the domestic pig derived from them was still a very coarse, slow maturing, and unprofitable animal. The French and Germans, as compared with the English, have made but little improvement in the breeds of pigs, and many of the animals to be found upon the Continent are very much like the old English hog, bony, tall, gaunt, wiry-haired, and slow to fatten.'

By the early eighteenth century there was plenty of variation in British pigs. Some were communal scavengers; some were cottage pigs; some lived well on the by-products of dairy and arable farming. Gradually regional types developed, differing in the carriage of their ears, the length of their bodies and legs, and their skin and coat colours. There was a broad tendency for white pigs to be favoured in the north (the famous Yorkshire pigs, for example), black or black-and-white pigs in the south, and spotted and blotched multi-coloured pigs of red, black and sand in between.

Although pigs had played a role in medieval agriculture, they declined sharply with the loss of woodland habitat in which to rootle and with the great increase in the planting of arable crops – no farmer wanted a pig rootling in his barley fields, and pigs suddenly found themselves being confined and controlled. They became mainly cottage livestock, kept in a sty and fed on whatever waste the cottager could find, and becoming a vital resource for the cottage larder in winter. Cottagers migrating to towns in search of work sometimes took their pigs with them, to scavenge in the streets during the day, and in the eighteenth century pig-racing in the streets was a popular spectator sport. One farmer used to drive into town with four large hogs drawing his carriage. In Scotland there were isolated reports of cows and sows being yoked together with horses for ploughing. Elsewhere pigs were used to hunt for truffles, and even in one or two cases to act as a gamekeeper's 'gun-pig' to point to and retrieve game.

Some farmers did continue to keep pig herds in areas where there were surpluses from their own cereal and bean crops, or where a local dairy or brewery produced useful by-products for feeding to pigs, but in 1776 the agriculturalist John Mills wrote: 'Of all the quadrupeds that we know, at least certainly of all those that come under the husbandman's care, the Hog appears to be the foulest, the most brutish, and the most apt to commit waste wherever it goes. The defects of its figure seem to influence its dispositions: all its ways are gross, all its inclinations are filthy, and all its sensations concentrate in a furious lust, and so eager a gluttony, that it devours indiscriminately whatever comes its way.' However, attitudes to the pig were about to change.

Quite early in the eighteenth century the first known imports of Asian pigs arrived in small numbers, and from the 1770s in greater numbers. These mainly **Cantonese** and **Siamese** pigs fattened much more quickly (with a

The long-legged 'Old Irish Greyhound' pig, illustrated in about 1850 – tall, bony, heavy-eared, coarse-haired and said at the time to be an intermediate form between the Wild Boar and the domesticated pig. Similar pigs could still be found in Galway in those days and were described as a 'white' breed by some of the old writers. Note the tusks, and the wattles dangling under the jaw.

Imported Cantonese sow. The original Chinese pigs were mainly imported between 1770 and 1780 from Canton; they seem to have been mostly white, but occasionally black or pied. Most of the importations were of boars (rather than sows), used to improve the old Berkshire pig.

A highly influential Siamese sow, imported via Singapore by Messrs Dugdale of Manchester, painted as a three-year-old by William Shiels and reproduced by Professor Low in 1842. She has a typically straight oriental tail. Her litter (by a half-bred Chinese boar) reveals a mixture of colours and coat patterns.

The famous Neapolitan breed was first imported by Lord Western (1767–1844) of Rivenhall, Essex, from his Italian travels and described by him as 'a breed of very peculiar and valuable qualities, the flavour of the meat being excellent, and the disposition to fatten on the smallest quantity of food unrivalled'. It had no bristles, but sparse silky hair and fine skin. Similar pigs were imported from Malta. This boar and sow were the property of the Rt Hon. Earl Spencer and had been imported from Naples by the Hon. Captain Spencer.

thick layer of blubber) and could be slaughtered at nine months old or less; their meat was more delicate, and they were very docile. They had characteristically pendulous bellies almost touching the ground, dipped backs, short little legs, short little upright ears, very little hair, fine bones, and often dished or squashed faces. Some slate-coloured copper-skinned **Neapolitan** pigs also arrived, larger and more elegant than the Asian pigs but with Asian blood, a good flavour to the meat and better mothering abilities.

Pig-breeding suddenly became fashionable, and countless new breeds were created, especially by the upper classes and particularly for the show-ring.

Big English sows were bred to the 'highly refined' Chinese boars and thereafter selective breeding for form and quality gradually established the different British breeds. By the early nineteenth century, pigs of every conceivable colour, coat pattern and type could be found in most parts of England. The passion for Chinese pigs lasted for nearly a century, and their impact on British pigs was substantial and radical. No more was heard of them, however, after about 1860, until in the 1980s Britain's big pig-breeding companies once again began to import Chinese pigs to improve the fecundity of the national herd.

To write of the 'British' pig is perhaps misleading. Wales had plenty of pigs, but in Scotland pig meat was never very popular and pig-breeding was much less important than elsewhere in the British Isles: there seem to have been no true Scottish breeds. In Ireland pigs were plentiful, and some very good pigs were produced over the centuries. But the breeds that became famous all over the world, exported in huge numbers during the nineteenth and early twentieth centuries in particular, were largely English breeds. Here is their story.

Good old-fashioned rustic pig of no particular breed.

21

COLOURED PIGS

THE ENGLISH MIDLANDS in the eighteenth century was the heartland for slouch-eared, splodgy-coloured pigs. There was, for example, a large hefty spotted curly-coated **Warwickshire** pig, a red-and-black or brindle **Shropshire** pig (as well as a better-quality Shropshire white), wire-coated black-and-white spotted pigs brought across the border from Wales, blue-and-white **Cheshire** pigs, red-and-white pigs in **Herefordshire**, and a mixture of white, coloured and spotted **Staffordshire** pigs that were either large and slouch-eared or smaller and prick-eared.

Another pig was widespread in the Midlands – and in Yorkshire, Devon, Norfolk and elsewhere – by the 1780s. This was the **Berkshire**, and soon almost any good-quality pig was being crossed with the Berkshire to improve the local type.

For a long while the old Berkshire was not uniform enough to be called a breed. It could be almost any colour: black-and-white, black-and-sandy, reddish brown with black spots, and so on. Some had large lop ears falling over the eyes; some had prick ears. Most had shortish legs, large bones, a long body and a great propensity to make fat – a considerable merit at the time. The type probably originated around Wantage, now in Oxfordshire but then in Berkshire.

In the 1820s the Berkshire was described as having a long and crooked snout, the muzzle turning upwards; it had heavy hanging ears and a long body and was of great size. Then Lord Barrington and others started to improve it, by crossing it with the Asian pigs, and it eventually became prick-eared, shorter in the body, finer-boned but still quite a large breed. The colour remained varied: in 1847 it was black-and-white and sandy spotted, or sandy or whitish brown, spotted regularly with dark brown or black spots. Some animals had the white 'points' (possibly from the Siamese) that later became the breed's trademark. In due course the main colour was black. At one stage its face became very dished: show-ring breeders liked the snubby oriental look, and snubbiness also deterred rooting.

Opposite: British Saddlebacks, a very useful outdoor breed that is increasing in popularity.

This 'British Boar' was originally painted by Edwin Landseer in 1818. The boar belonged to C. C. Western Esq., MP for Essex.

The Berkshire became the favourite pig of the nineteenth century, widely used to improve other British pigs and to produce crossbreds for slaughter, but its popularity declined during the twentieth century until it became a rare breed.

Today the Berkshire is an attractive-looking prick-eared breed – said to have a good sense of humour – and has improved considerably to meet more modern tastes in recent years. Although it is black-haired (with touches of white on the legs, face and tail switch), its carcass 'dresses out' white, and the flesh is fine, with a high proportion of lean to fat; and it matures early. The sows make excellent easy-going mothers with plenty of milk for their litters, which average nine piglets.

Classic painting by William Shiels of a Berkshire (reproduced in David Low's series of livestock illustrations in 1842). This sow was bred in Warwickshire by Mr Loud of Mackstockmill. Her colouring is not seen in today's Berkshire, except for the white socks and tassel.

Above and left:
Evolution of the
Berkshire improved
with Chinese blood.
(Top left) The old
lop-eared spotted
type; (above) an
1845 cross between
Chinese boar and
old Berkshire sow –
note the prick ears
and the black, white
and sandy spotted
coat; and (left) the
'Improved
Berkshire', now
black with white
touches (ideally four
white feet, a white
spot between the
eyes, and a few
white hairs behind
each shoulder), in
this case developed
by Mr Sadler, of
Bentham, near
Cricklade
(Wiltshire), from
stock originally
obtained from Lord
Barrington (died
1829), the great
improver of the old
Berkshire.

Left: Peasant pigs,
one a Berkshire-
like pricked-eared
black and the other
a spotted lop (the
latter not to be
confused with the
spaniel-like dog on
the far right).

The Berkshire, now black with white points, was the most successful and widely used of the improved pigs in nineteenth-century Britain, and popular with the aristocracy. It is now a rare breed but is sometimes used as a sire on commercial sows to produce piglets for outdoor rearing.

Closely related to the original Berkshire was the **Tamworth**, initially a red-and-black Staffordshire pig, which is said to have been one of the few British pigs (if not the only one) *not* to be crossed with Asian pigs in the eighteenth and nineteenth centuries. While the Berkshire breeders did their best to lose the redness of their breed, the Tamworth breeders did their best to retain it, and by the 1820s most Tamworths were a whole-coloured deep red or mahogany (skin as well as coat). The colour softened during the nineteenth century and is now a golden red, but the breed retained a long snout, upright ears, a long lean body that made superb bacon, and legs that apparently enabled it to jump over any barrier.

The adventurous golden red Tamworth is one of Britain's most striking pig breeds, with a distinct lack of Chinese influence.

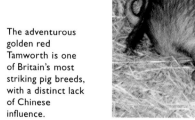

The unusual colour of the Tamworth has led to endless speculation about its origins: an Indian jungle boar imported by Sir Francis Lawley; a West Indian boar crossed by Sir Robert Peel with an Irish Grazier sow; a much earlier red pig from Barbados crossed with local Wiltshire pigs; a West African Guinea hog (descended from Portuguese pigs), and so on. But such a colour does not require an exotic source, even though this is now Britain's only native red breed.

The Tamworth is a talkative pig of considerable character and is another of the rare breeds: it reached a peak of popularity in the early 1950s, but numbers had dropped close to extinction by 1973. Today the numbers are increasing – and the breed received an unexpected boost to its popularity when the 'Tamworth Two' escaped pigs became nationally famous in the British media. Its fine meat has a distinctive flavour for bacon, pork and ham. Tamworth sows cross well with the commercial white breeds, but the traditional cross has always been with the Berkshire. It is an excellent outdoor pig, resistant to sunburn and thoroughly hardy.

In the early 1970s Joe Henson of the Cotswold Farm Park, the co-founder of the Rare Breeds Survival Trust, crossed a Wild Boar with a Tamworth sow for the sake of a group of archaeologists wanting livestock for a reconstructed Iron Age village. The resulting piglets were striped, like those of the Wild Boar, and Henson continued to breed his **Iron Age** hybrids, selecting for the darkest and most docile ones.

The 'Iron Age' hybrid is a cross between Wild Boar and Tamworth. The piglets are born with wild-type stripes.

The colour of the old Berkshires and Tamworths is reflected today in the **Oxford Sandy and Black** (OSB), a breed with a chequered and sometimes disputed history. It might have derived originally from an old Oxford Dairy pig crossed by the Marquess of Blandford with his black

Neapolitan boar and a later cross with Essex blacks to create the Improved Oxford. The cottager's 'plum pudding' black-and-white spotted Oxford was possibly from the Tamworth crossed with the Neapolitan. As there were assorted spotted pigs all over the Midlands in the eighteenth and nineteenth centuries, the OSB might have developed from any of them. As a breed it seems to have become extinct, or nearly extinct, at least twice, most recently in the 1960s. Unfortunately, it is easy to produce a litter of piglets with OSB colouring simply by crossing Berkshire and Tamworth with each other or with other breeds (and quite a few farm parks like to experiment with such crosses for the joy of the multi-coloured piglets that result), but usually such offspring have prick ears: the genuine OSB does not, and the original breed also had a white 'tapir' stripe or face blaze.

Original Oxford Sandy and Black photographed on Norman Boseley's farm at Sarsden Lodge in 1959, showing ear carriage and white stripe. In 1881 Joseph Harris referred only to an 'Improved Oxfordshire' mentioned by Sidney in 1850, scarce even then and jet black all over, based on crossbreeding Berkshire and Improved Essex. But a sandy-coloured old Oxford, 'approaching the Tamworth type, though not of that breed', was mentioned in the 1940s as having existed in the county 'until recent times'.

The OSB, with its attractive red coat blotched with black, is now rare but rapidly increasing. It is a good-natured and hardy pig, producing good meat as a purebred and also crossing well with commercial whites.

Another pig that originated from the Tamworth, probably crossed with Berkshire and Gloucestershire Old Spots, was the very pretty and well-named **Dorset Gold Tip**. By 1955, however, there was only one registered boar, and the breed soon became extinct.

Left: The striking Dorset Gold Tip was a light Tamworth-red with black markings and characteristic sun-glinting gold tips to its bristles.

The Midlands spotted pigs included yet another favourite British breed: the **Gloucestershire Old Spots** or Orchard pig. The old Gloucester was of the typical Old English slouch-eared type, dirty yellowish-white in colour, and traditionally it had wattles (fleshy tassels) dangling from its jowls. The Old Spots type developed in the Berkeley Vale (probably through crossing with the old Berkshire), where it thrived on whey and windfalls. The breed had a vague history until a breed society was formed in 1913, when selected spotty pigs were taken into the herdbook. It was unpretentious but a good farmer's pig, producing heavy hams and good bacon, prolific, docile behind its lop ears, a grazer, and very hardy. The breed society marketed the breed magnificently, and there was a huge

Opposite Bottom: The attractively marked Oxford Sandy and Black is a popular choice for farm parks and children's farms.

Gloucestershire Old Spots sows and their litters in the days when the emphasis was on the spots.

Above: The GOS, fondly known as the Orchard pig, is popular with smallholders and increasingly finding a role as a sire in commercial free-range herds.

Above right: America's Poland China is black with white points but was originally spotted. The 'Spotted Poland China' shown here from the 1940s resulted from the importation of a pair of GOS from England. In recent years the name of the Spotted Poland China has been shortened to 'Spotted Swine' or simply 'Spots'.

boom in spotted pigs, but the bubble burst in the 1940s and it became rare. In the early 1970s there were only thirteen registered Gloucestershire Old Spots boars and it is still a rare breed, though with higher numbers than other rare breeds. Today it is thoroughly hardy, a good grazing pig, very happy in a paddock with supplementary apples, roots and whey, and can be reared to produce both pork and bacon. The sows are noted for being docile and long-lived mothers. Because of butchers' prejudice against colour, the number and size of the target-like spots on the coat were reduced considerably, often to only one or two, but more spots are acceptable now.

In the south of England the coloured pigs tended to be black, or black with white belts. They are now represented by the Large Black and the British Saddleback, both of which are the result of merging West Country breeds with East Anglian ones.

The **Large Black** originated from the Old English type crossed with the copper-skinned black Neapolitan and with Asian pigs. In the days of big sailing ships, livestock was carried on board to supply the crew with food, and it is recounted that when a ship landed at Plymouth some little black Chinese pigs that had escaped the pork barrel were bought by local farmers, who bred them with their own big Cornish pigs (possibly French in origin) to create a black Devon. A similar story is told on the other side of the country: shipboard Chinese pigs landed on the east coast and were crossed with local pigs there to create, ultimately, a **Small Black** combining the **Black Essex**, the **Black Suffolk** and the slate-blue roly-poly improved **Black Dorset**, each with its own intriguing history.

The Dorset pigs had been bred from a pair of Somerset pigs 'of a breed said to have been sent from Turkey' – possibly a mixture of Wild Boar and Neapolitan, subsequently crossed with Chinese. (The 'Turkey' pig was probably a Siamese imported from Tunkey, or Tonquin, now northern Vietnam.) Prize Dorsets, fed on cream and treacle, became so obese that pigmen placed wooden pillows under their snouts to avoid self-suffocation.

Left: Lord Western, the Essex squire and the original importer of Neapolitan pigs, created his own very inbred Neapolitan and used the breed to improve the local Essex pigs (described in the 1830s as having a 'roach back, long legs, sharp head and restless disposition'). He probably also incorporated black Sussex and Berkshire breeding to make his all-black 'Lord Western Essex' shown here. The new Black Essex went on to contribute to many other improved English breeds, including the Improved Essex created by Western's tenant farmer, Fisher Hobbs.

Below: These two nineteenth-century pigs, belonging to the Duchess of Hamilton at Easton Park, Suffolk, were described in the 1890s as 'Suffolk or Small Black'.

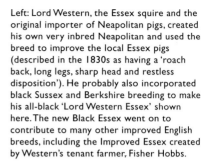

The Improved Essex came from Squire Western of Felix Hall, the original importer of the black Neapolitan pigs from Italy. He crossed his imports on big Essex sows to create his **Essex Half-black** (black one end, white the other), and his tenants used the cross to create the Improved Essex as a small fat black pig giving top-quality pork.

Large black or 'blue' pigs were found right across southern England from Cornwall through Sussex to Kent. At the very end of the nineteenth century all of them were brought together to form the **Large Black**, which also absorbed the Small Black of East Anglia. The Large Black was a very popular and very large bacon breed in the 1920s but is rare today. Its 'mealy' black colour protects it in hot climates, and it is

Grossly overweight Improved Dorsets, illustrated in 1874, with chin pillows to help them to breathe.

Right and middle:
The Large Black
(still known as the
'Cornwall' in parts
of Europe) remains
large and mealy
black, with lop ears
and deep full hams.
It is a much finer
and longer pig than
it used to be but is
now rare.

Below right:
Described as a
'Somersetshire',
this belted sow
belonged to the
Dowager Duchess
of Devonshire
(living in
Eastbourne). The
breed was said in
1908 to be 'the
next suppliant for
fame and a herd-
book all to itself'
but was the victim
of jealousy
between its
breeders and
those of the Large
Black. The
Somersetshire was
blue-and-white,
usually belted or
'sheeted', and was
compared to the
American
Hampshire.

thoroughly hardy in its own country as a handsome and gentle outdoor grazer.

Some of the old Essex pigs were described as 'sheeted', meaning black with a white band over the shoulders that also extended to cover the forelegs. This pattern is now known as 'saddleback', and the **Essex** developed into a smart-looking saddleback breed with four white socks and a white tail tassel. It was good for both pork and bacon, and it was once suggested that it was a cross between Large White and Berkshire.

An Essex in the show ring. Note the four white legs that distinguish the Essex from the Wessex, in which only the forelegs are white.

Below and bottom: Wessex Saddleback. (Below) Royal Show champions in 1932 belonging to Mr Holloway of West Lavington, Wiltshire. (Bottom) Herd featured in 1921 sale particulars.

In the west of England there was another saddleback but with only two white feet – the **Wessex Saddleback**, probably developed by crossing dark spotted New Forest or Hampshire bacon pigs with the neighbouring massive black Sussex. The ears of the Wessex tended to pitch forward more than those of the Essex, which was an altogether finer and lighter pig. The Wessex's saddle was not as broad as that of the Essex, but the qualities of the two became similar and they were combined as the **British Saddleback**. It is still possible to discern the Essex and the Wessex within the combined breed. The breed is increasingly popular for outdoor systems and on organic pig farms.

WHITE PIGS

MANY LOCAL white breeds have disappeared over the years or been absorbed by the few that remain. The old **White Leicester** was possibly the result of crossing pigs bred by one of the pioneers of livestock breeding, Robert Bakewell (1725–95), and originating from white, light spotted or mahogany local Leicester pigs crossed with a black boar, or from white Chinese, or it might simply have arisen from breeding local pigs with the soon to be famous Middle White or Small White from Yorkshire. Despite its name, the White Leicester of the latter half of the nineteenth century was sometimes categorised as 'brown' or was described as light with black or brown spots.

In Lincolnshire the white pigs grew remarkably thick curly coats as protection against the weather. These fleecy, broad-backed, flop-eared **Lincolnshire Curly Coats** were lovely big pigs of character, great

The 'stupendious' Yorkshire Hog, exhibited to the public for his immense size in 1809. He was bred by Benjamin Rowley of Red House, near Doncaster, and fed by Joseph Hudson on the estate of Colonel Thomas Charles Beaumont, MP, of Bretton Hall. At four years old, the hog was 9 feet 10 inches long and 12½ hands high and weighed 1,344 pounds.

favourites with cottagers and smallholders, but the breed was extinct by the early 1970s.

The **Cheshire** pigs – white, blue-and-white or black-and-white – were enormous, with long hanging ears, narrow curvy backs, big heads and long bony legs. Cully (1807) described a particular Cheshire pig that was 9 feet 8 inches long and 4 feet 6 inches high and weighed 1,410 pounds.

One of Britain's biggest breeds in the mid nineteenth century was the **Yorkshire and Lincolnshire**, said to 'exceed in weight that of a moderately grown Scotch Ox'. It was not beautiful, having very long legs, a very long body, poor conformation and coarse flabby flesh, but it was probably the progenitor of the most famous breed in the world, the Large White or, as it is still known in many countries, the **Large Yorkshire**. The old Yorkshires were huge and hungry but were hardy and prolific and produced good bacon. They were improved with this and that – Berkshire, Black Essex, Neapolitan and so on – creating a black-and-white mixture. Then the prick-eared White Leicester was used on them early in the nineteenth century. Still the breeders, earls and dukes among them, continued with their improvements, and perhaps the most successful was a Keighley weaver, Joseph Tuley, in the middle of that century. Tuley pigs were exported worldwide and could fetch as much as £1,000 for one sow.

Gradually the Large Yorkshire became white and absorbed other county whites: the **Large White** was recognised in 1868, and its first herdbook was published in 1884.

Large White sows on pasture in the early years of the twentieth century.

Now the most numerous pig in Britain (and probably in the world), the prick-eared and hearty Large White, with its slightly dished profile, is well adapted to intensive farming systems for bacon and pork.

There was, briefly, a small **Yorkshire Blue-and-White**, or Bilsdale Blue, which became extinct in the late 1950s. But creating blue-and-white bacon pigs by crossing Large White boars on Large Black sows was common practice for a long time.

For many years the term 'Yorkshire' included a wide range of sizes within the breed. They evolved into the Large White, Middle White and Small White, with an increasing Chinese influence as they became smaller. A prize-

Large White, famous worldwide and probably the most populous of any breed. In many countries it is still known as the Large Yorkshire. Outdoor boars tend to be rather hairier than the indoor gentlemen used by the big commercial companies.

winning sow, Lady Kate, was accompanied in its railway truck to all the shows by the devoted wife of its owner, who would sit beside the sow and 'reckon up on her fingers its thirteen crosses from the Chinese'. A boar named Arch Trespasser was exhibited at the Royal Show as a Small White at one year old, as a Middle White at two and as a Large White at three.

The prick-eared **Small White**, said to have originated from shipboard white Chinese pigs landing at Plymouth and Bristol in the eighteenth century, became oddly popular with well-known breeders of Shorthorn cattle as well as the toy of aristocrats: many a lord and gentleman renamed it as a breed after himself, and even Prince Albert exhibited some. His pigs were 'magnificent specimens of fat and cleanliness' but also 'absolutely gasping for breath', their snouts being so pug-like. Not surprisingly, the Small White was extinct by about the time of the First World War.

The **Middle White** was a more sensible and less extreme pig, and by the 1930s it had become famous as the London Porker and suckling pig. It was widely exported, especially to Japan (where it was a royal favourite and

Small White boar and sow in the early 1900s. The Small White was described in the 1890s as 'not so much a tenant's pig as a landlord's or a gentleman's pig', with a snout 'much contracted and quite turned up ... like the nose of a pug dog'.

Middle White, evolved from the old Large Yorkshire crossed with Chinese pigs in the nineteenth century. At some stages in its history the snout became so squashed in profile that breathing could be a problem.

The common European Landrace pig of the Celtic type that later developed into several Landrace breeds in Europe.

where a shrine was erected to a Middle White boar). It was about the size of a Berkshire but much lighter-boned. Today the Middle White is smaller than most British pigs and is a rare breed, valued for early maturity as a pork pig and crossing well for pork and bacon with other breeds. It is instantly recognisable by its squashed face.

The Large White is one of Britain's three main commercial breeds; the other two whites are the British Landrace and the Welsh. The lop-eared **Landrace** was introduced into the United Kingdom in 1949 from Sweden and deliberately bred for British conditions and markets. The sows are good, prolific and docile mothers and are often crossed with Large White.

The **Welsh** has had a varied history. It is claimed that pigs came into Wales in Viking times with the Celts who took refuge there (mass pig migrations are recorded in the *Maginogion* saga). Welsh pigs in later history were typical of the Old English lop-ears, usually yellowish, sometimes with black spotting; they and the Cornish pigs were described as 'wolf-shaped' in the eighteenth century. There were a few primitive brown pigs, similar to those in Scotland described in 1872 as 'an alligator mounted on stilts'.

The highly commercial modern Landrace is a typical lop-eared white European pig, quick to grow, quite long in the body, and with a lean carcass. Its lop ears and long snout are some of the features that distinguish the Landrace from the Large White. The British version evolved from Swedish pigs imported in the 1950s.

Rather coarse Welsh pigs came over the English border in large numbers during the nineteenth century. In 1907 Youatt described them as chiefly white and 'very much of the razor-backed, coarse-haired, slow-maturing kind'. Then the improvers began to work on them and developed a lop-eared white Welsh pig. By 1946 it had become similar to the famous Danish Landrace; it was a large breed, long in the body, with a slightly dished face like the Large White but with the ears forward over the eyes and tending to meet towards the tips. It was a dual-purpose type, good for bacon or for pork. By 1949 the new breed was very low in numbers (only thirty-three registered boars) but it was boosted with Swedish Landrace blood, and by the early 1980s it had become Britain's third most numerous breed.

The modern Welsh is essentially similar to the post-war breed. It is a long-bodied lean white lop-eared pig of the Landrace type. It is able to be reared outside as well as indoors, with a high-quality carcass. Its numbers are declining and it is classified by the Rare Breeds Survival Trust as 'vulnerable'.

At one stage in its development, the Welsh Pig Society combined with the societies of the

Old Glamorgan and the Long White Lop-Eared of south-west England. The latter pig is an old and practical breed that did not stray far from its West Country roots, though today there are herds as far away as Scotland. Threatened with extinction by 1970, it rallied during the 1980s and is now a minor breed known as the **British Lop**. Britain's rarest pig breed, it has had its own breed society in Cornwall since 1918 and is still sometimes called the Cornish White or the Devon Lop. Tavistock was the main centre

The Old Cornish White, now known as the British Lop.

for its development. It is probably closely related to the old Welsh pigs and to the large White Ulster, and perhaps to the old Cumberland (extinct by 1960). Well suited to the small-scale farms of the south-west, the Lop is thoroughly hardy and can be reared economically on stubble, pasture and woodland. It can be a porker; it also produces good streaky bacon and crosses well with the commercial white breeds for porkers and baconers. One of its drawbacks is that it is broadly similar in appearance to the British Landrace and the Welsh, and so some people interested in rare breeds tend to pass it by.

The now extinct **Large White Ulster** was Ireland's only truly commercial breed. It resembled a Large White but was finer and silkier, with long ears falling forwards over the eyes rather than being erect. It was a grazer and bacon breed with a short history: the herdbook was opened in 1908, but the last boar was licensed in 1956. Its thin skin, desired by bacon curers, unfortunately bruised too easily in transport.

In the past Ireland had produced some interesting breeds or types, the best of which was the **Irish Grazier**, a term that covered a range of pigs with thin skins and coats of various colours or white, and of various types and sizes. They included a meatier, long-bodied, full-hammed type with erect ears. Many were exported to the United States in the 1830s and contributed there to the Poland China and Chester White.

British Lop, a large West Country breed, sometimes mistaken for a Landrace but a very rare breed. It is long in the body, white in colour, with long thin lop ears inclining over the face, and long silky white hair. Its conformation is similar to that of the Large Black.

NEW PIGS

THE PIGS of Britain continue to evolve, and in the commercial sector various imported breeds have made their mark in the continual quest for improvement. Ironically, pigs that are superficially (in colour) similar to some of the British rare breeds, and which originally owed their development to English pigs, have been brought over from the United States to improve the British gene pool. These American breeds include the red Duroc, the belted Hampshire and the semi-lop Chester White.

The **Duroc**, originally known as the Duroc-Jersey or simply the Red Hog, may or may not have had an infusion of red pigs from the African Guinea coast (quite possible in slave-trading days) or from Portuguese and Spanish pigs in the early days of colonisation, but the old Berkshire is another strong candidate for its colour (Tamworth blood was tried but apparently the results were so unsatisfactory that the offspring were 'discarded'). The modern Duroc ranges in colour from a light golden yellow through auburn to a dark mahogany red, with no colour markings, and it has a thick winter coat (though it can look almost bald in summer). Its ears droop forwards. The British Duroc has evolved since the 1980s and is used mainly as a female line to produce crossbreds for outdoor pig farms.

The **Hampshire**, an important breed worldwide, is claimed to have originated from the Old English white-belted black hog. Importations into America in the late 1820s were said to have been via ports in the English county of Hampshire, hence the name; the type was also known as the McKay after the man who shipped them into the United States. Others suggest the type originated in the New Forest in Hampshire. The breed developed mainly in Kentucky, where it was known as the Thin-Rind and became very popular with butchers. In 1904 belted hogs with names as varied as Thin-Rind, McKay, McGee and Ring Middle were formally named the Hampshire. It was first imported into the United Kingdom in 1968 and there is now an established British Hampshire. The colouring and pattern are similar to that of the British Saddleback but the ears are erect rather than lop.

Opposite:
Britain's ubiquitous Large White is increasingly being crossbred with imported American breeds.

The red American Duroc is one of the most numerous pig breeds in the world. The British Duroc, a useful terminal sire on Large White and Landrace sows, has been developed from the American breed, which was first imported into the UK from Canada in 1968 and later from the US and Denmark.

The **Chester White**'s background seems to have included white pigs originally from Yorkshire and Lincolnshire bred from around 1818 to an imported white boar, said to be either a Bedfordshire or a Cumberland (two English counties with many miles between them). It is a good outdoor white pig with semi-lop ears.

Chinese pigs are once again influencing British breeders. In the 1980s one of the big pig-breeding companies began to use the **Meishan** to increase the prolificacy of its commercial line. The Meishan is one of the black, wrinkle-skinned, pot-bellied Taihu group of breeds from the Lower Changjiang (Yangtze)

The American Hampshire, a major black-and-white saddleback in its home country and an important breed worldwide.

The belted British Hampshire was developed from the American Hampshire (first imported into the UK in 1968) for crossing with commercial white sows and for creating new hybrids. In contrast to the British Saddleback, the Hampshire has prick ears.

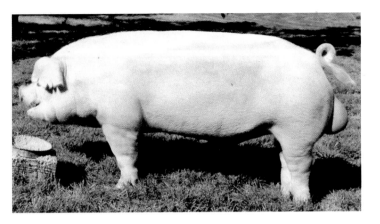

Chester White, promoted as a 'good milker', and the sows cross well with Duroc boars.

Basin. The average number of the third litter is sixteen; it is not uncommon for them to have twenty piglets at a time, and they can produce two litters a year. The **Meidam** is an example of a commercial purebred dam line developed in Yorkshire from the Meishan: the Meidam has sixteen teats (compared with fourteen in the British Landrace). The new dam line, like others developed from Chinese breeds in Britain, has a much better carcass than the original Meishan.

The Chinese Meishan has wrinkled skin and heavy great ears drooping down like a spaniel's. Its belly is so low slung that it almost brushes the ground in a pregnant sow.

Pig-breeding
companies used
the Meishan to
increase fecundity
in British
commercial lines,
initially by
breeding imported
Meishan sows to
Large White boars.
The crossbred
sows retained the
Meishan's dipped
back in the early
stages of the
programme.

Other British pig-breeding companies followed suit with the Meishan and also with breeds such as the **Jinhua**, or 'black-at-both-ends' pig (white, with a black head and a round black patch like a target on its rear). The Jinhua is a pretty pig, rather endearing in its dumpiness, with a low belly in typical Chinese fashion, and ears hanging over its eyes. Females can be mated as young as three months old. It is of the same type as the black-and-white Cantonese, which was the original Chinese improver in Britain in the eighteenth century.

The small Jinhua was imported from China in the late twentieth century to improve prolificacy in British commercial breeds.

More recently it has been a two-way trade. In 2000, for example, British pig companies were exporting white piglets and sows in their thousands to China in joint ventures worth billions of pounds, to meet a big surge in demand for lean pork in urban parts of the Far East. The Chinese expressed admiration of British breeds of pigs (and poultry) and of British breeding technology, which was exported along with the animals.

Vietnamese Pot-bellied pigs became quite popular as pets and on farm parks, but were originally introduced to the UK as laboratory animals.

The low-slung bellies and extremely early sexual maturity of the Chinese breeds are also characteristic of the related **Vietnamese Pot-bellied**, a dwarf type that became fashionable in the United Kingdom (and in the United States, imported via Canada) in the 1980s and 1990s as a pet, until it grew rather larger and more demanding than its fond owners had anticipated. Most of those in Britain are of the black, wrinkle-skinned breed known simply as Í, from the Red River delta, but some of the pot-bellies may be spotted or solid white.

Other pot-bellies seen in Britain include the **Mong Cai**, which comes from the same area as the old Siamese or 'Turkey' (Tonquin) pigs that sailed

The miniature Froxfield Pygmy was bred for the laboratories from crosses between the Vietnamese Pot-bellied and the Yucatan. (Above) Vietnamese x Yucatan sow with piglets. (Right) Froxfield Pygmy.

to England in the eighteenth century. The Mong Cai is white with a black head and black patches elsewhere on the body (often on the rump or as a saddle); it has a white snout and sometimes a white forehead star.

New breeds have also been created for laboratory work, where small size is a useful characteristic. For example, the **Froxfield Pygmy** was bred in England by crossing the Yucatan Miniature (originally bred at Colorado State University from Yucatan local pigs selected for small size) with the Vietnamese Pot-bellied, resulting in a small, manageable miniature that was initially a coloured or spotted pig but was gradually selected for a white coat.

The newest imported breed to catch the eye of British smallholders and owners of pet pigs is the delightful little **Kunekune** or Maori pig from New Zealand. Its name means 'round and fat' and its origins are obscure. The pigs seem to have a touch of Chinese about them and may have arrived with the Maoris from Polynesia; but it is equally likely that they were of European origin, accompanying early seafarers or settlers. They are small; the adults are about 24–30 (60–76 cm) inches tall and 28–36 inches (70–90 cm) long, and they have small tassels or wattles (known as *pire pire*) under the jaw. The ears vary from prick to lop. Colours range from cream or ginger to brown or black and spotted, and the coat is either long and curly or short and straight. It is for their nature that they are becoming popular: they are gentle, placid, easy-going, sociable and friendly, and easy enough to keep on grass.

Above and left: The little Maori pig, or Kunekune, has replaced the Vietnamese Pot-bellied as a favourite at farm parks and for some smallholders. It often has wattles under the throat and its coat colours, patterns and hair texture vary considerably.

Pennywell Mini pig, bred for its small size and amenable nature as a pet, well suited to being handled by children. The Kunekune features in its ancestry.

A large farm park near Buckfastleigh in Devon has created its own **Pennywell** breed of pet miniature pigs from a mixture that includes Iron Age and Wild Boar, with a top cross of Kunekune, selecting for small size (a 'pocket-size pig ideal for children and adults to have on their laps'), character, liveweight gain, fecundity, food conversion ratio, litter size, temperament and also a wide colour range that produces ginger, brown, black, mixed-spotted, white-spotted and 'Dalmatian'-spotted coats.

The Mangalitza was once widespread in central Europe. A typical lard pig, illustrated here in the late 1930s, it was almost spherical, late to mature and produced only three to seven in a litter, but could cope with hot, dry Hungarian summers and severe winters – it had a thick curly coat. Because of coat similarities, the Lincolnshire Curly Coat was imported by Hungarian pig farmers in the nineteenth century and crossed with the Mangalitza.

Now rare in its native Hungary, the Mangalitza has recently been imported by an English farmer (with a soft spot for the extinct Lincolnshire Curly Coat) who offers pig-keeping courses to smallholders. (Left) Fleecy coat in close-up. (Below left) 'Swallow-bellied' coat pattern.

In 2006 a pig rare breeds centre (Pig Paradise Farm) in Wiltshire imported more than a dozen unrelated individuals of an old Hungarian curly-coated lard breed, the splendidly woolly though rather primitive **Mangalitza**. The Lincolnshire Curly Coat, which became extinct in Britain in the early 1970s, had been exported to Hungary

Man riding a hog, illustrated by engraver Thomas Bewick (1753–1828) in his *History of British Birds.*

more than a century earlier and did well in local harsh winters; it was to some extent crossed with the Mangalitza and the offspring were known as the Lincolitsa in the 1920s. The Mangalitza is nearly extinct in its homeland but the Wiltshire farm eventually found some breeding stock in three different colour lines – blonde, red and 'swallow-bellied' (with a pale underside) – and has started its own British breeding programme for these fleecy pigs.

One of the newest pigs in Britain is the oldest: the **Wild Boar**, long since extinct in the wild but now re-imported from continental Europe for its low-cholesterol meat. It is being farmed commercially in several places. Farmers require a special licence and substantial fencing (some animals have already escaped into the countryside and caused a few problems), but Wild Boar are perfectly happy in woodland, whether confined or not, and they soon adapt to become as manageable as any other British pigs.

The genetic diversity of other wild pig species is being investigated at the Roslin Institute in Scotland. *Sus scrofa*, the Wild Boar, is only one of several related pig species. Others include Asian species such as *Sus barbatus* (bearded pig) and *S. verrucosus* (Javanese warty pig), which might have genes that would be useful for commercial pig breeders. Stretching the relationship a little, the Institute is also looking at Africa's warthog (*Phacochoerus africanus*) and bush pigs (*Potamochoerus* spp.) and even the extraordinary-looking tusky Indonesian babirusa (*Babyrousa babyrussa*), to see whether they too might contribute to the improvement of the domestic pig. It is some nineteen million years since these species diverged from the ancestor they shared in common with *Sus scrofa*, the species that is believed to be the source of all today's domesticated pigs worldwide over the past nine thousand years. But recently it has been shown that the Asian and European races of Wild Boar separated into different populations half a million years ago – hence the noticeable differences even now between Asian and European domestic pigs.

Studies are in hand to see which of today's European breeds originally derived some of their genes from Asia as well as Europe. Even that very British breed, the Large White or Yorkshire, seems to have genetic material that could only have been inherited long ago from Asian sows.

It is fitting that Asian and European pig breeders are deliberately combining their breeds in the twenty-first century, especially at a time when European and American commercial breeds are beginning to dominate China's production systems so that the diversity of China's own breeds is under threat. Genetic diversity is an insurance against future needs and it remains important to conserve the rare breeds that we do have. They became rare because markets and farming systems changed, but there might be unimagined future changes in which their breed qualities will once again be needed.

Gloucester Old Spots piglets at a farm park, well used to children.

FURTHER READING

Harris, Joseph. *Harris on the Pig: Breeding, Rearing, Management, and Improvement.* Orange Judd Company, New York, 1881.

Long, James. *The Book of the Pig: Its Selection, Breeding, Feeding, and Management.* L. Upcott Gill, London, and Charles Scribner's Sons, New York, second edition (revised) 1906.

Porter, Valerie. *Practical Rare Breeds.* Pelham, 1987.

Porter, Valerie. *Pigs: A Handbook to the Breeds of the World.* Helm Information, 1992. Illustrated by Jake Tebbit.

Porter, Valerie (editor). *Mason's World Dictionary of Livestock Breeds, Types and Varieties.* CABI Publishing, fifth edition 2002.

Wallace, Robert. *Farm Live Stock of Great Britain.* Crosby Lockwood & Son, London, third edition 1893.

Litter of Gloucester Old Spots piglets.

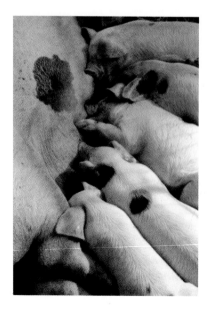

Video

Porter, Val. *In Pursuit of Pigs: A Celebration of Breeds.* (Narrated by Paul Heiney.) Farming Press Videos (Miller Freeman plc, 2 Wharfedale Road, Ipswich IP1 4LG), 1997.

USEFUL ADDRESSES

British Kune Kune Pig Society: Secretary, Hannah Smith.
 Telephone: 01348 840098.
 Website: www.britishkunekunesociety.org.uk
British Lop Pig Society: Secretary, Frank Miller.
 Telephone: 01948 880243.
 Website: www.britishloppig.org.uk
British Pig Association:
 Telephone: 01223 845100.
 Website: www.britishpigs.org.uk.
 For full details of commercial breeds (Large White, British Landrace,
 Welsh) and for minor breeds (Berkshire, British Duroc, British
 Hampshire, British Saddleback, Gloucestershire Old Spots, Large Black,
 Mangalitza, Middle White, Oxford Sandy and Black, Tamworth).
Oxford Sandy and Black Pig Society: Mrs Heather Royle, Lower Coombe
 Farm, Blandford Road, Coombe Bissett, Salisbury SP5 4LJ.
 Telephone: 01722 718263.
 Website: www.oxfordsandypigs.co.uk
Rare Breeds Survival Trust: National Agricultural Centre, Stoneleigh Park,
 Kenilworth, Warwickshire CV8 2LG.
 Telephone: 024 7669 6551.
 Website: www.rbst.org.uk
 For full details of rare/minor breeds and their breeders' clubs,
 including Berkshire, British Saddleback, Gloucestershire Old Spots,
 Large Black, Middle White, Tamworth. Also details of rare breed farm
 parks open to the public, and butchers supplying rare breed meat.
 The quarterly journal *The Ark* includes pig contacts.
The Real Boar Company: Cotswolds. Telephone: 01249 782861.
 Website: www.therealboar.co.uk
 Rearing free-range Wild Boar; processing and selling Wild Boar meat.
Traditional Breeds Meat Marketing Company: Richard Lutwyche.
 Website: www.tbmm.co.uk.

PLACES TO VISIT

A selection of centres and breeders with several rare pig breeds (contact the Rare Breeds Survival Trust for more). Telephone for opening times before planning a visit.

Cotswold Farm Park, Guiting Power, Cheltenham, Gloucestershire GL54 5UG.
   Telephone: 01451 850307.
   Website: www.cotswoldfarmpark.co.uk
   Most of the rare breeds.
Croxteth Home Farm, Croxteth Hall and Country Park, Liverpool L12 0HB.
   Telephone: 0151 233 6910.
   Website: www.croxteth.co.uk
   Saddleback, Tamworth, Berkshire, Large Black.
Cruckley Animal Farm, Foston-on-the-Wolds, Driffield, East Yorkshire YO25 8BS.
   Telephone: 01262 488337.
   Website: www.cruckley.co.uk
   Most of the rare breeds.
Newham Grange Leisure Farm, Coulby Newham, Middlesbrough TS8 0TE.
   Telephone: 01642 300202.
   Website: www.middlesbrough.gov.uk
   Several rare breeds.
Pennywell Farm Activity and Wildlife Centre, Buckfastleigh, Devon TQ11 0LT.
   Telephone: 01364 642023.
   Website: www.pennywellfarm.co.uk
   Own breed of miniature Pennywell pigs.
Pig Paradise Farm, Wiltshire.
   Telephone: 01380 818677.
   Website: www.pigparadise.com
   Pig-keeping courses; stock for sale: Berkshire, British Lop, British Saddleback, Gloucestershire Old Spots, Kune Kune, Large Black, Middle White, Oxford Sandy and Black, Tamworth; Mangalitza breeding programme. Not open to general public.

A Wild Boar family
tasting freedom.

Sherwood Forest Farm Park, Lamb Pens Farm, Edwinstowe, Mansfield,
  Nottinghamshire NG21 9HL.
  Telephone: 01623 823558.
  Several rare pig breeds.
South of England Rare Breeds Centre, Highlands Farm, Woodchurch, Ashford,
  Kent TN26 3RJ.
  Telephone: 01233 861493.
  Website: www.rarebreeds.org.uk
  Most rare pig breeds.
Tatton Home Farm, Tatton Park, Knutsford, Cheshire WA16 6QN.
  Telephone: 01625 534431.
  Website: www.tattonpark.org.uk
  Most rare pig breeds.
Temple Newsam Home Farm, Temple Newsam Estate, Leeds LS15 0AD.
  Telephone: 0113 264 5535.
  Website: www.leeds.gov.uk/templenewsam
  Several rare pig breeds.
Wimpole Home Farm, Wimpole Hall, Arrington, Royston, Hertfordshire
  SG8 0BW.
  Telephone: 01223 208987.
  Website: www.wimpole.org
  Several rare pig breeds.

To see prize pigs, visit the agricultural shows, many of which have pig
classes. For show dates and places, contact the *Association of Show and
Agricultural Organisations*, The Showground, Shepton Mallet, Somerset BA4
6QN. Telephone: 01749 822200.

INDEX